CAUSE OF DEATH

CAUSE OF DEATH

A Psychological Autopsy of Jesus

STEVE RUSH

Contents

Also by the Author

Introduction

As a young boy I became fascinated with crime scenes, victims of violent crimes, and an investigator's ability to solve them. One major question rose in my mind: How did investigators know what to look for at any given crime scene? I could not figure out how detectives determined what part of the crime scene had evidentiary value and what did not. Then, I wondered how investigators were able to identify the person responsible for the act of violence based solely on what was discovered at the crime scene.

My inquisitiveness, along with education and years of experience finally resulted in the answers to my questions. Like a puzzle, pieces were finally identified and rightfully placed.

I understood more about the process as I grew older and became educated in investigative techniques and procedures involving complex crime scenes, police investigations, and the fascinating world of forensic medicine. As each section fit into place, I saw clarity in the total picture.

Understanding the process of investigation, in theory, is as simple as watching thespians solve crimes on *CSI* or some other crime drama. In reality, the process is complex and often long-term.

Assimilation of critical information hopefully will lead to the ultimate goal. That goal is truth. In our business, goals are accomplished by application of the Scientific Method—a six-step process identified as an ever-expanding and self-correcting body of knowledge.

Once the problem or question has been identified, data must be collected to resolve the problem and provide an answer. This may involve experimentation, research, examination of scene photographs and reports and any other information collected during the investigation. Data collected may be in the form of both physical and testimonial evidence. The end result will identify evidence and place it in one of two categories: circumstantial or real.

Looking for identifying factors as motive and opportunity along with use of hypotheses aid investigators in analyzing data collected in order to find a solution to the problem or answer additional questions based on interpretation of information available at the time. Every hypothesis must be tested against any collected evidence for support. The hypothesis is either supported or rejected based on available evidence.

One crucial piece of the evidentiary puzzle at any scene involving violence is the presence of blood.

Blood tells the story.

Life of all flesh is in the blood. From a criminal-justice perspective, blood is defined as, "A vital, complex biological fluid, containing red blood cells, which is present in vertebrates and may be shed during accidental, intentional, and/or criminal acts."

The total blood volume in a human body is about five liters.

The body cannot survive without it. Severe, rapid blood loss leads to shock and loss of life.

The sight of blood loss, especially in large volumes, causes investigators to become naturally suspicious. From a detective's point of view, the presence of blood in specific patterns and unusual places is sure evidence of some questionable event having occurred.

Analysis of blood stains has proven to be a valuable tool in crime-scene reconstruction. A close analysis of blood stain patterns provides insight into the event. If the incident involved some violent act, blood stain analysis often leads to the identification of weapon type used in the incident by the perpetrator.

Blood is identified as the physical evidence referenced in the first recorded murder. Cain committed the murder of his brother, Abel, whose blood implicated Cain as documented in Genesis 4:10. *"The voice of your brother's blood cries out to Me from the ground."*

The physical evidence of Abel's blood spoke in a more convincing voice than did Cain's denial of the crime.

The cellular components and plasma of blood result in easily identifiable spatter patterns when blood is shed or lost. This is due to the surface tension of blood: the molecular cohesiveness that causes the surface of a liquid to resist penetration. Blood spatter patterns are easily identified to the trained eye. The various types of blood spatter fall under one of four categories: free falling blood; projected blood; cast off blood; or transfer blood.

After having investigated numerous crime scenes and completed a study in blood spatter analysis, initial scene investigation and reconstruction of scenes took on a whole new perspective for me.

Blood spatter analysis as an investigative tool caused my curiosity to peak when I thought about what otherwise might have been in certain matters. Many unanswered questions to some of the most heinous crimes most probably would have been answered if only the investigating agencies had been afforded some knowledge of blood spatter interpretation and its application.

One crime in particular is the assassination of President John F. Kennedy.

Forget what witnesses indicated they saw on the grassy knoll or the resounding blasts of a firearm heard on that dreadful November day in 1963. An examination of the presidential vehicle by an expert in blood spatter analysis, along with the presence of other forensic evidence, would within scientific certainty and probability have resolved the question of which direction the fatal shots came from. Nevertheless, because of the lack of spatter analysis and probable contaminated or lost evidence on or about the presidential vehicle, a thorough reconstruction of the event in these terms was not possible.

The cohesive nature of blood results in predictable and reproducible patterns. Therefore, replication for reconstruction purposes end in reliable results.

When valuable evidence is lost, we are left with speculations of what might have been.

In this study, we will examine the events surrounding the death of Jesus Christ. Although we are not afforded the opportunity to examine the evidence first-hand, historical and biblical records exist to place Jesus in specific places at specific times. Therefore, based on accounts of what we know happened and the types of injuries Jesus sustained, attempts to reconstruct the scene and identify the type of death He died has a more promising outcome.

The physical effects of Christ began in the Garden of Gethsemane. There in Gethsemane we have the first record of Jesus' blood loss.

Stress began to take its toll on Him due to anticipation and dread of what was to be.

Blood tells the story in Gethsemane as single drops of blood fell to the earth, spattering on the rocky soil. As this phenomenon continued, one by one they left behind evidence as to their origin.

Later, Jesus faced emotional torture directed at Him from a variety of sources, including mockery at the hands of the Roman soldiers, rejection by His own people, and denial by one of His closest disciples. Even one of His chosen disciples betrayed Him for thirty pieces of silver.

Jesus faced the prospect of physical pain, including beatings and scourging followed by torture and death on a cross.

Even the stress of facing torture of a spiritual nature was not to escape Him. He willingly took the cup filled with sin and accepted accountability for fallen humanity, then cried out to His Father; "My God! My God! Why have you forsaken me?"

Laden with the weight upon Him, Jesus went to Gethsemane on the Mount of Olives for a time of prayer.

According to one description, Jesus began to passively lose blood. *"His sweat became like great drops of blood falling down to the ground."* (Luke 22:44b)

If indeed Jesus did sweat blood, drops of blood would have fallen to the ground. Blood would have stained His clothing. Blood would transfer to objects He would have touched.

Assuredly, some evidence of blood loss would be seen as He left Gethsemane at which time He was met by the throng of government and religious officials led by the one who betrayed Him.

Arrested by soldiers, Jesus was taken into the city of Jerusalem where He faced the mockery of a trial at the hands of the Sanhedrin and Caiaphas, the High Priest. There in the presence of the Sanhedrin Jesus sustained the first physical and emotional trauma inflicted in the form of simple battery and mockery.

Blood tells the story when the next morning Jesus was taken before the Roman government, where at some time that morning Pilate condemned Him to scourging and crucifixion.

Stripped of His clothing, Jesus had His hands affixed to a post above His head. A Roman soldier administered harsh blows to the exposed flesh using a whip know as a flagellum or flagrum (cat-o-nine-tails). This scourging led to appreciable blood loss.

Blood tells the story from Pilate's Hall where the trail of blood led down Via Dolorosa on the way to Golgotha. Somewhere between Pilate's Hall and Golgotha Jesus, either too weak to continue or possibly having collapsed under the weight of the cross, was assisted by a man identified as Simon of Cyrene. From there the trail continued on to Golgotha where Jesus, the Son of God was crucified.

Large spikes were driven through Jesus' flesh, securing His body to the cross. The cross was raised, suspending the wounded and bleeding Jesus between heaven and earth in an open display for all to see. His blood continued to flow, stained the cross and dropped to the earth.

Here at Golgotha the trail of blood ended.

It was the end of the beginning.

I

Agony in Gethsemane

The Garden of Gethsemane and Mount of Olives, located on the northeast side of Jerusalem, Israel, was an area frequented by Jesus Christ during His life on earth. On numerous occasions Jesus would remove himself from the company of His disciples and go to the Mount of Olives for a time of prayer.

For the benefit of this investigation and reconstruction of events, our focus will begin with the Last Supper and Jesus' last known visit to Gethsemane. During this last visit Jesus dealt with the emotional aspects of His impending torture and death. There, Jesus relinquished His will and agreed to take the cup of sin for humanity. In doing so, the agony and grief He encountered was so great that His emotional state was described as "exceedingly sorrowful," "troubled and deeply distressed," and "being in agony."

The depths of understanding the mystery of someone encountering agony exceeding what an ordinary person might encounter appears cloudy at first glance. Comprehension will allow us to go only as far as what we have witnessed either in our lives or in the lives of others close to us.

In a mental picture of the Last Supper, I visualized Jesus sitting at the table with the twelve disciples. Arrangements were made for the group to have what would be their last meal together. For three years they served the one they called Master. Their journey as a cohesive group of followers as they knew it was about to come to an end. Life for each was about to take a dramatic turn.

One of them even pre-arranged to betray Jesus.

Judas Iscariot met with the chief priests before the time of this last meal and contracted with them to betray Jesus. All Judas needed to do was wait for the right time to carry out his end of the contract.

This collusion is documented in Matthew chapter twenty-six. *Then one of the twelve, called Judas Iscariot, went to the chief priest and said, "What are you willing to give me if I deliver Him to you?" And they counted out to him thirty pieces of silver. So from that time he sought opportunity to betray him.*

During the meal, Jesus told His disciples about His upcoming betrayal. Based on their reaction they wondered among themselves which one of His chosen would be capable of betraying Him. Who among Jesus' chosen disciples could possibly want to commit an act like this? At the same time the question "Is it I?" among them reveals an impression of possible doubt within each of their minds concerning their own potential.

Now as they sat and ate, Jesus said, "Assuredly, I say to you, one of you who eats with Me will betray Me." (Mark 14:18)

"But behold, the hand of My betrayer is with Me on the table." (Luke 22:21)

"Most assuredly, I say to you, one of you will betray Me." (John 13:21b)

The apostle John leaned over and rested his head against Jesus. Simon Peter knew John was the disciple Jesus loved. Therefore, Peter coaxed John into asking Jesus to identify the betrayer.

Then the disciples looked at one another, perplexed about whom He spoke. Now there was leaning on Jesus' bosom one of His disciples, whom Jesus loved. Simon Peter therefore motioned to him to ask who it was of whom He spoke. Then, leaning back on Jesus' breast, he said to Him, "Lord, who is it?" Jesus answered, "It is he to whom I shall give a piece of bread when I have dipped it." And having dipped the bread, He gave it to Judas Iscariot, the son of Simon. (John 13:22-26)

As Jesus sat holding the Holy Grail, Judas dipped in the cup and then took leave of them. The time had come. The Son of God was about to be betrayed.

The days preceding this last supper were no doubt filled with anguish because of Jesus' anticipation and endless thoughts of things He was about to encounter. In all probability, Jesus experienced sleepless nights and meal-less days as a result of the anxiety that enveloped Him. This emotional course corresponds with Isaiah's description of the "man of sorrows" and the reference to he who was "acquainted with grief."

Anticipatory pain and suffering are not things even Jesus would have taken lightly. Jesus was mindful of the impending destruction of His physical body. In less than twenty-four hours

Roman soldiers would scourge and crucify Him. This known fact of Roman crucifixion instilled terror in its victims.

Jesus was aware of the overwhelming pain He would suffer as a result of this physical destruction as well as His having to take on humanity's sin and become the only sacrifice required to provide atonement. The price for redemption was high and Jesus knew no one else could possibly pay it.

From personal experience we often find it difficult to relax and fall asleep when our minds are racing with thoughts and fear of what is to be, especially when the outcome is known. Even eating during times like these is almost impossible because of nausea and accompanying symptoms associated with these stressful situations. Nevertheless, faith in God is always the beacon of hope we have to lead us through each one.

Jesus was no different. Although He is God, when Jesus walked on the earth He was as human as you and me. Jesus faced the same type of anxieties we face. His dread of having to drink the cup of sin, being nailed to a cross and relinquish His life as a sacrificial lamb was as real for Him as it would be for any of us if we were in the same situation. His burden exceeded heaviness. His agony was real. His suffering exceeded anything we might ever experience. Jesus suffered in every way imaginable in addition to becoming the very thing He despised—sin.

With these things weighing on Him, Jesus and those with Him departed the upper room. They left the city and walked across the brook Kidron to the Mount of Olives and Gethsemane. Arriving there, He left the company of the disciples for a time of prayer.

When Jesus had spoken these words, He went out with His disciples over the Brook Kidron, where there was a garden, which He and His disciples entered. (John 18:1)

Then Jesus came with them to a place called Gethsemane, and said to the disciples, "Sit here while I go and pray over there." (Matthew 26:36)

Coming out, He went to the Mount of Olives, as He was accustomed, and His disciples also followed Him. (Luke 22:39)

I imagine Jesus knelt on the bare ground near one of the many olive trees and turned His face toward the night sky. The sky was dark. The full moon illuminated the garden and His face as He began to pray. The twinkling stars emitted their light with every ounce of energy within them in an attempt to reach down to earth and shine on the face of their Creator.

While Jesus prayed, the sound of His sobs could be heard throughout the garden. I wonder what He was feeling. By this time He probably felt abandoned and maybe forgotten. He was —and would continue to be—despised and rejected by His own people. At that very moment Jesus was being betrayed by one of the ones chosen to serve with Him exclusively in His ministry.

And Judas, who betrayed Him, also knew the place; for Jesus often met there with His disciples. Then Judas, having received a detachment of troops, and officers from the chief priests and Pharisees, came there with lanterns, torches, and weapons. (John 18:2-3)

Even when Jesus arrived at the garden, He asked three of His most trusted disciples to watch for Him while He prayed, only to return an hour later to find them sleeping. This happened not once but three times. Did they not care? Were they not concerned about what was happening to their Master? Is it possible that disappointment or displeasure was obvious in His voice when He said to them, "sleep on"? I would surmise both to be true.

The cup Jesus was about to drink from was filled to the brim with sin. To accomplish the task, He was about to undertake meant He would have to empty the contents of the cup onto Himself. He would have to become the very thing He abhorred. He would have to become sin. It was not only the weight of the cross that would bring Jesus to His knees. The weight of our sin accomplished that hours before Jesus ever reached the cross.

Anticipatory Pain and Suffering

Jesus was soon to experience physical pain in its extremist form. He was about to be beaten, scourged, and finally, crucified. Mere anticipation of the immense physical pain Jesus faced would have caused an enormous amount of stress on even the strongest of wills.

Although we are not able to see into the mind of another person to state with certainty what they are experiencing or opine on their "state of mind," Jesus' emotional state in Gethsemane is identified using a "psychological autopsy" based on His actions, which are described by some as follows:

Matthew (Matthew 26:37) described Jesus as "sorrowful and deeply distressed." The Greek word "ademoneo" is also translated "very heavy" and is said to imply a "separation." It is when a person is depressed and almost overwhelmed with sorrow or burden of mind. Matthew continues (verse 38) by saying He was "exceedingly sorrowful," "perilupos," which describes a person who is encompassed, encircled, overwhelmed, and severely grieved.

Mark used the phrase (Mark 14:33) "troubled and deeply distressed," "ekthanbeo," describing someone who is greatly amazed or astonished from terror or distress of mind. In verse thirty-four Jesus said, "My soul is exceedingly sorrowful, even to death."

Luke uses the words "being in agony" (Luke 22:44). The word agony is from the Greek word "agonia." It implies a sense of combat; used to refer to the trembling excitement and anxiety produced by fear or tension before a fight. This is not a fear prompting someone to flee but one that trembles in the face of the issue yet continues on until the end.

Sweating Blood

And being in agony, He prayed more earnestly. Then His sweat became like great drops of blood falling down to the ground. (Luke 22:44)

Does this description by Luke imply that Jesus sweated blood while He prayed? Is it possible for a person to sweat blood? If so, what would cause this phenomenon to occur? Are there any unusual circumstances involved? Are there any documented cases where a person was known to have perspired blood?

First, is it possible for a human to sweat blood? The answer is yes. Sweating blood can and has occurred. This rare phenomenon is identified as hematidrosis. A thorough search of medical literature substantiates this phenomenon, which is a condition resulting in the excretion of blood in the sweat. Tiny capillaries in the sweat glands rupture, mixing blood with perspiration. This occurs under conditions of great emotional stress. The most frequent inciting causes have been identified as acute fear and intense mental contemplation.

A May 11, 1918, British Medical Journal article, *A Case of Haematidrosis*, describes a young girl who, having a fear of air raids during World War I, developed this condition after a gas explosion in her neighbor's home.

Johann S. von Grafenberg, in his 1585 article, *Observations Medicae de Capite Humano*, tells of a Catholic nun who was so terrified after being threatened by sword-beating soldiers that she bled from every part of her body and died.

In the 1996 Journal of Medicine article, *Blood, Sweat, and Fear: A Classification of Hematidrosis,* seventy-six cases of this phenomenon were studied.

According to Luke, a known physician in that era, Jesus experienced this rare phenomenon. Suffering from hematidrosis does not imply that physical pain accompanies any aspect of the process. Nevertheless, this process results in marked weakness and causes the skin to become extremely sensitive to touch, thereby making the impending physical injuries Jesus suffered even more painful.

In his sermon, *The Agony in Gethsemane,* Charles Spurgeon stated, "The woe that broke our Savior's spirit, the great and fathomless ocean of inexpressible anguish which dashed over the Savior's soul when he died, is so inconceivable, that I must not venture far, lest I be accused of a vain attempt to express the unutterable; but this I will say, the very spray from that great tempestuous deep, as it fell on Christ, baptized Him in bloody sweat."

Jesus' emotional state as a result of the above description while in Gethsemane would have to be considered inconceivably beyond limits of extreme for this corresponding bloody sweat to occur.

Emotions

A physician in neuropathology defined emotion as "a way of feeling and a way of acting. It may be defined as a tendency of an organism toward or away from an object, accompanied by notable bodily alteration. There is an element of motivation; an impulsion to action and an element of alertness, a hyperawareness or vividness of mental processes."

From the above definition we are able to identify the aspects of cognition, expression, and experience. Perception and possible evaluation of any given situation takes place before any emotion occurs. Once a thing is perceived emotion is expressed either outwardly, inwardly, or both. The experience that follows depends on how a person feels based on their perception and expressed reaction.

Jesus perceived and evaluated His dilemma before His emotional reaction was evident to any of those around Him. He became engaged in a battle with Satan and His own desire to live.

A psychiatrist described Jesus' response to what He faced in Gethsemane: "The temptation for me to run was strong. I could feel one of my greatest battles beginning. I wanted to retaliate, but I couldn't. I had to take the evil of the world—death itself in all its forms including my father's judgment—and not give it back. I had to take it to the grave. I knew horrible destruction was mine if I accepted it, and fear engulfed me. The next day was already so real to me that my sorrow and grief were too much. It was literally killing me. Once, when I opened my eyes, I looked down at the ground where I was kneeling and saw a pool of blood. As I wiped the sweat pouring from my forehead,

I realized I was in so much agony that I was sweating drops of blood...the grief and the suffering wouldn't let up."

During His ministry, Jesus often told His disciples what was going to happen; that He would freely give His life as a sacrifice for humanity; that He would willingly pay the ransomed price to redeem them from destruction and finish the work He came here to do.

"Behold, we are going up to Jerusalem, and the Son of Man will be betrayed to the chief priests and to the scribes; and they will condemn Him to death and deliver Him to the Gentiles; and they will mock Him, and scourge Him, and spit on Him, and kill Him." (Mark 10:33-34)

"Behold we are going up to Jerusalem, and the Son of Man will be betrayed to the chief priests and to the scribes; and they will condemn Him to death, and deliver Him to the Gentiles to mock and to scourge and to crucify." (Matthew 20:18-19)

One forewarning was given two days before His betrayal.

"You know that after two days is the Passover, and the Son of Man will be delivered up to be crucified." (Matthew 26:2)

When Jesus finally reached the place where He planned to pray His emotional reactions were expressed outwardly in the form of tears and lamentation. At the same time His emotional reaction inwardly caused a form of vascular changes which resulted in bloody sweat.

Jesus' body and His clothing were wet and stained with bloody sweat. His visual appearance was different than it was when He entered Gethsemane.

The ground underneath where He prayed exhibited bloodstain evidence of His presence.

I have no doubt about the amount of mental anguish Jesus

must have sensed in Gethsemane. The presence of hematidrosis is undisputable evidence of the "exceedingly sorrowful" condition He was described to have been in as a result of His anticipatory pain and suffering. Imagine how Jesus felt that evening in Gethsemane. Consider the impending "cup" of sin about to be poured out on Him in addition to the temptation by Satan, rejection by His Father, and the cruel torture and death which neared.

Remember, while Jesus was on the earth He was as human as you and me. If Jesus had decided not to take the cup and drink it what would have become of us? Everything He went through was for the benefit of humanity and was an expression of the immeasurable love He has for us. Therefore, He knew what had to be done to save humanity from the destructive power of sin. He had to become the sacrificial substitute in our stead. He had to come under the curse once placed on humanity and taste death for every individual. This He knew and willingly stated "not my will."

The evidence found in Gethsemane emphatically indicates that in Jesus we have a person of unusual will and determination. Jesus was determined to accomplish the task He came here to do at any and all cost to Himself. What we see in Jesus during this time in the Garden of Gethsemane is a son submitting verbally and physically to the will of a father. His words and actions displayed love in its truest form to the extent that He relinquished any sense of self-preservation. He overcame the temptation to fight for survival and every destructive emotion He had to deal with.

He knew the time was near. He accepted the task long

before His arrival on earth. Now He was prepared to see it to completion.

2

Arrest and the Sanhedrin

"Sleep on," Jesus told His disciples after His time of prayer. "The time has come." Light from torches and lanterns once viewed a distance away drew closer to the garden. Jesus knew Judas was on his way to Gethsemane with soldiers and a host of others to affect His arrest.

Other than the name Jesus, the most recognized name among Christians and non-Christians alike would most probably be that of Judas Iscariot. The name Judas Iscariot has a well-known and documented association with betrayal and deceit. What a legacy to leave behind. The name Judas has been the brunt of many serious implications and jokes alike since that historical day. The person of Judas Iscariot has been branded unlike any other.

On this night, Judas brought with him a mixture of soldiers, servants of the High Priest and others to the Garden of Gethsemane. This band of the captain and officers of the Jews were armed with swords and staves. Although Judas may not have known it at the time, their plan all along was to put Jesus to death.

Then the chief priests, the scribes, and the elders of the people assembled at the palace of the high priest, who was called Caiaphas, and plotted to take Jesus by trickery and kill Him. (Matthew 26:3)

After two days it was the Passover and the Feast of Unleavened Bread. And the chief priests and the scribes sought how they might take Him by trickery and put Him to death. (Mark 14:1)

Then the detachment of troops and the captain and the officers of the Jews arrested Jesus and bound Him. And they led Him away to Annas first, for he was the father-in-law of Caiaphas who was high priest that year. Now it was Caiaphas who advised the Jews that it was expedient that one man should die for the people. (John 18:12-14)

Judas kept his end of the deal and led the multitude to Gethsemane for the sole purpose of having Jesus arrested. The sign of a single kiss from Judas identified Jesus to the mob.

And while He was still speaking, behold, Judas, one of the twelve, with a great multitude with swords and clubs, came from the chief priests and elders of the people. Now His betrayer had given them a sign, saying, "Whomever I kiss, He is the One; seize Him." Immediately he went up to Jesus and said, "Greetings, Rabbi!" and kissed Him. (Matthew 26:47-49)

And immediately, while He was yet speaking, Judas, one of the twelve, with a great multitude with swords and clubs, came from the chief priests and the scribes and the elders. Now His betrayer had given them a signal, saying, "Whomever I kiss, He is the One; seize Him and lead Him away safely." (Mark 14:43-44)

On that night instead of attempting to hide or flee, Jesus went out to meet His accusers. Jesus' true character was exemplified before them all. When He asked them who they were looking for they said, "Jesus." Jesus answered, "I am He."

What they observed in Jesus that night was nothing like what they expected. The time had come and He, with resolve and calm countenance, spoke in a freely surrendering manner. His demeanor left the crowd in a total state of confusion and disbelief. That night Jesus would not "escape" his captors. The courage Jesus displayed in His willingness to face the "enemy" echoed the strength and determination necessary to complete the work He had to do. No longer would he remove himself from those who sought to do Him harm. He knew it was time and he was prepared to face all they were about to inflict upon his body.

The cost for the possession to be purchased was expensive. It was something Jesus wanted. When we see an item we desire, more often than not we inquire about the cost. We may desire the item but instead our minds become preoccupied with the number on the price tag. At that moment we may not see beyond its price. The desire to possess an item a lot of times may not outweigh the price on the tag. Therefore, we often leave instead of paying the price and taking possession of what fulfills our desire.

Jesus looked beyond the price and saw the prize. He could see beyond the pain and torture into the future. What He saw was our future. For the reason of love, He would allow His body to be torn. He would allow His dignity to be brought to an open shame. In an expression of true love, He was willing to give His life for the purchased possession—humanity.

The antagonistic band followed Judas to the Garden of Gethsemane and took Jesus forcefully even though He offered no resistance.

From Gethsemane they took Jesus to Caiaphas, the High Priest.

What transpired over the next several hours was clearly unrelenting brutality. The force used to affect Jesus' arrest and His treatment while in custody exceeded that which was necessary.

In today's judicial system, charges would be brought against those who physically abused Jesus on that night. Nevertheless, Jesus made no complaint against anyone. No attorney dared to step forward to represent Him. There was no offer of any defense before the Sanhedrin on His behalf. No one voiced any complaint about violation of His civil rights.

Jesus stood alone before Caiaphas and members of the Sanhedrin. He had seen the system in action and knew their intentions. Did He not have a right to representation? Did He not have a right to a fair trial? "The tribunal about to condemn Him, it must not be forgotten, was not a legal 'court,' but simply a self-constituted 'Committee of Public safety' extemporized by the excited Temple authorities and Rabbis, like the Vigilance Committees of America; with a Jewish Fouquier Tinville for President, in the person of the Sadducee Caiaphas. Knowing the illegality of their proceedings, they could only venture to propose the framing an indictment to lie before Pilate, and trust to their violence for extorting a condemnation from him.

The hierarchy was masters of form, and knew how to honor the appearance of justice while mocking the reality."

The Sanhedrin was made up of seventy-one members, composed of the religious Sadducees and Pharisees; chief priests, scribes and elders, with the High Priest filling the role of presiding judge. This assembly religious leaders acted as both judge and jury. If one were to have personal knowledge or interest in the outcome of any given matter, he was required to recuse himself from the case. Members were required to have a character of unimpeachable integrity.

Before and during trial the presumption of innocence ruled. Like our judicial system, any accused appearing before the Sanhedrin was to be considered "innocent until proven guilty."

According to *Mishnah Sanhedrin 4.1,* capital cases are to be held during the day and the verdict must also be reached during the day. If the verdict was to acquit it may be the same day. If the verdict was guilty, they must wait until the next day.

It was the Sanhedrin's responsibility to hear the matter against the accused according to the laws of evidence and if witnesses were to be called, no less than two witnesses were permitted to testify and the witnesses, if only two were present, must agree on every point. "No case could be moved to trial without witnesses because Jewish law was based on the testimony of others than the defendant."

Although witnesses appearing before the Sanhedrin were not administered an oath, it was felt that the Commandment *"You shall not bear false witness"* was sufficient to deter perjury. In addition, other deterrents to perjury included subjection to the death penalty if committed during a capital case and, if the accused in a capital case was convicted, the witness found to be guilty of perjury was required to attend the execution. Under this provision witnesses were careful to choose their words cautiously.

Witnesses had to be proven. Circumstantial and hearsay evidence were not admissible. According to the *Mishnah Sanhedrin* 5.1, "They used to prove witnesses with these inquiries: In what week of years? In what year? In what month? On what day? In what hour? In what place?" Based on the record, witnesses present to testify against Jesus were not proven.

Since the Sanhedrin was considered the highest authority why then would they have deviated—as indicated—from Jewish law by conduction of an improper trial? The answer may be found in the history between the Sadducees and Pharisees—who were part of the Sanhedrin—and Jesus.

In a Sanhedrin trial, as is stated in our 5th Amendment, the accused could not be required to testify. This lack of testimony offered by the accused was not to be held against them. "So long as the prisoner remained silent, the Great Sanhedrin had to acquit." Therefore, the Sanhedrin had no right to "demand" any answer from Jesus and could, according to the rule of law, only offer a defendant an opportunity to speak on his behalf. Any question directed at the defendant would be considered out of order and, therefore, inadmissible. Jesus knew Jewish law and knew what the consequences would be by answering any questions asked Him. Jesus willingly, knowing what was to be, condemned himself by answering, "I am." By speaking in answer to the High Priest's question, "Are you the Messiah, the Son of God?" Jesus admitted before the tribunal to being the Son of

God and, in the eyes of the Sanhedrin, condemned Himself. As a result, the Sanhedrin declared Jesus guilty of the charge of blasphemy, even though Jewish law would not allow a confession, believing a person's own words were unreliable.

Jesus never admitted to anything except His identity. Leaders of the tribunal began ranting and raving because of a simple answer to the question, "Who are you?" The directive, "Please state your name for this court" is normal procedure in any courtroom before an accused or any potential witness is allowed to offer testimony on any given matter before the court. When Jesus was arrested and taken to His so-called trial the High Priest could not get past His identity.

The most notorious criminals in history were never convicted on the basis of their identities. Convictions were and are handed down because of some criminal act(s) they were proven beyond reasonable doubt to have committed. Courts hold people responsible for their actions, not their identities or ancestry.

The Sanhedrin unlawfully condemned Jesus to death when the body of the council shouted, *"He is guilty of death"* (Matthew 26:66). They convicted and passed the death sentence on the same day as the trial—a direct violation of their own law (*Mishnah Sanhedrin 5.5*).

"The usual sentence was death in one of four ways: stoning, burning, decapitation or strangulation ... Each sentence of death was passed on, before execution, to the Roman Procurator. He usually affirmed it without question. If he felt so disposed, or if someone had pleaded the case with him privately, the Procurator might summon the accused and accusers and hear the case before his own judicial chair. If he reversed the verdict, the condemned was acquitted and the Sanhedrin was powerless. If he confirmed it, the Governor often forced the prisoner to undergo the Roman form of execution: crucifixion."

They took Jesus to Annas, father-in-law of Caiaphas. The trial before Annas was illegal. Annas possessed no right to conduct a trial.

The Sanhedrin trial of Jesus was without structural integrity. It was completely out of order. The Sanhedrin disobeyed their own laws to accomplish what they wanted. In doing so, they treated Jesus unjustly.

The Sanhedrin instigated charges against Jesus instead of investigating them. Jesus was not presumed innocent. He was compelled to testify, although He was not required by law to answer. Witnesses were sought who might be willing to testify to false claims and thereby would be granted immunity against perjury, but none were found. Finally, two witnesses were located who were willing to offer testimony, but even they could not agree on the facts of the case.

Before the night was over, Jesus was mocked, beaten, spit on, and condemned to death by the Sanhedrin who later admitted to Pilate, "It is not lawful for us to put a man to death."

3

Roman Trisl

"He delivered Him to the people on the day before the un-leavened bread, their feast. And they took the Lord and pushed Him as they ran, and said, 'Let us drag away the Son of God, having obtained power over Him.' And they clothed Him with purple, and set Him on the seat of judgment, saying 'Judge righteously, O king of Israel.' And then one of them brought a crown of thorns and put it on the head of the Lord. And others stood and spat in His eyes, and others smote His cheeks: others pricked Him with a reed; and some scourged Him, saying, 'With this honor let us honor the Son of God.'"

The Gospel of Peter

Confirmation of a Roman trial for Jesus, though sketchy, may be found in two historical accounts in addition to the writings of the gospels. These historical accounts were documented by Roman historian Tacitus and Jewish historian Flavius Josephus.

Tacitus wrote: "Christus, from whom the name had its origin, suffered the extreme penalty during the reign of Tiberius at the hands of one of our procurators, Pontius Pilate, and a most mischievous superstition, thus checked for the moment, again broke out not only in Judea, and the first source of the evil, but even in Rome, where all things hideous and shameful from every part of the world find their centre and become popular."

Flavius Josephus wrote: "About the same time, there lives Jesus, a wise man for he was a performer of marvelous feats and a teacher of such men who received the truth with pleasure. He attracted many Jews and many Greeks. He was called the Christ. Pilate sentenced him to die on the cross, having been urged to do so by the noblest of our citizens; but those who loved him at the first did not give up their affection for him. And the tribe of the Christians, who are named after him, have not disappeared to this day."

Jesus was taken to the Praetorium of the Fortress Antonia to appear before the Roman prefect of Judea, Pontius Pilate. His appearance before Pilate was for one purpose and one purpose only—to be put to death. Jesus was bound as though He was some notorious criminal and delivered to Pilate.

When morning came, all the chief priests and elders of the people plotted against Jesus to put Him to death. And when they had bound Him, they led Him away and delivered Him to Pontius Pilate the governor. (Matthew 27:1-2)

Immediately, in the morning, the chief priests held a consultation with the elders and scribes and the whole council; and they bound Jesus, led Him away, and delivered Him to Pilate. (Mark 15:1)

Arriving at the Fortress Antonia, the Jewish leaders addressed Pilate in an attempt to persuade him that Jesus posed an imminent threat to Roman authority since He was identified as "King of the Jews."

Pilate looked at Jesus and began to question Him about the accusations. "Are you a king?" The records specify Jesus' silence during the inquiry. Pilate's demeanor during this time suggests both astonishment and frustration as he continued his line of questioning. Pilate acted as though he could not believe Jesus kept silent when asked about the accusations against Him.

While He was being accused by the chief priests and elders, He answered nothing. Then Pilate said to Him, "Do you not hear how many things they testify against You?" But He answered him not a word, so that the governor marveled greatly. (Matthew 27:12-14)

Jesus did not appear to be a threat to Pilate even though a few days earlier Jesus rode into Jerusalem and was greeted by some who hailed Him as Messiah. They honored Jesus as they would a king. They lined his pathway with palm branches in His honor (John 12:12-13). Nevertheless, Jesus made no official declaration.

After only a short time, Pilate was satisfied Jesus posed no threat to either his position as prefect or to the Roman Empire. Jesus' demeanor proved neither forceful nor condescending. No doubt the information implicating Jesus as a threat to the Roman Empire came from the ones seeking Jesus' demise—the religious leaders of the Jewish nation.

During the proceedings, Pilate asked the question, "What is truth?" In asking, what was Pilate really hoping to learn from

Jesus? Whatever it may have been the Roman prefect seemed satisfied about Jesus' innocence. Pilate addressed the crowd and Jesus' accusers; "*I find no fault in him at all.*" Was any fault to be found in Jesus? The Roman ruler did not seem to think so; even to the extent that he finished his statement with "at all." None.

According to the record in John's gospel chapter nineteen, Pilate repeated his proclamation of innocence again and again. Three times Pilate said, "I find no fault in Him." Pilate discovered nothing in Jesus' character, His actions, or from testimony offered by anyone present to convince him of anything worthy of conviction for any crime; and even less convincing of a death sentence.

Pilate expressed willingness to release Jesus and indicated his intention to the waiting crowd. Accusations against Jesus continued from the crowd. When Pilate heard claims of Jesus being from Galilee, and realizing Herod was in Jerusalem at the time, he passed Jesus off to Herod.

They were the more fierce, saying, "He stirs up the people, teaching throughout all Judea, beginning from Galilee to this place." When Pilate heard of Galilee,*he asked if the Man were a Galilean. And as soon as he knew that He belonged to Herod's jurisdiction, he sent Him to Herod, who was also in Jerusalem at that time.* (Luke 23:5-7)

"The old palace of the Asmoneans, in which Antipas lodged, was a short way from Pilate's splendid official residence. It lay a few streets off, to the north-east, within the same old city wall, on the slope of Zion, the leveled crest of which was occupied by the vast palace of Herod, now the Roman headquarters."

Having heard about many miracles Jesus was said to have performed among the people, Herod hoped to see Jesus perform some miracle in his presence. When Jesus would not respond as Herod had wished, and since Herod was most likely humiliated by Jesus' silence, he and others present began to exhibit cruel disdain for Jesus. Herod and his men of war mocked Him.

The chief priest and scribes continued their accusations against Jesus.

After a time of interrogation and ridicule, Herod returned Jesus to Pilate.

Now when Herod saw Jesus, he was exceedingly glad; for he had desired for a long time to see Him, because he had heard many things about Him, and he hoped to see some miracle done by Him. Then he questioned Him with many words, but He answered him nothing. And the chief priests and scribes stood and vehemently accused Him. Then Herod, with his men of war, treated Him with contempt and mocked Him, arrayed Him in a gorgeous robe, and sent Him back to Pilate. (Luke 23:8-11)

After receiving word from Herod concerning Jesus, Pilate decided to have Jesus scourged and released. The puzzling thing is, if Pilate believed Jesus to be innocent and knew Herod held no charge against Him why would he have Jesus scourged while fully intending to release Him?

Scourging was a legal preliminary to Roman execution. Scourging prior to crucifixion averted a lengthy stay on the cross. I wonder if Pilate already had crucifixion in mind before sending Jesus off to be scourged. By threatening Jesus with scourging Pilate's actions may have been his attempt to coerce Jesus into a confession. Maybe Pilate expected Jesus to cower to the threat of scourging or the scourging itself and confess to the offenses He was accused of.

Sensitivity from the hematidrosis (sweating blood) had already affected Jesus' skin. Now Jesus had to endure torture of a scourge at the hands of a Roman lictor.

Roman guards lead Jesus to the whipping post where He was stripped of His clothing and His hands secured to the upper portion of the post. The bloodstained garment Jesus wore in the Garden of Gethsemane was in all probability the garment removed here.

When lictors scourged an individual with a flagrum it was with the intention of inflicting extreme pain to the point of death. The flagrum used by the Romans to scourge was a whip made of leather strips. The strips varied in length. Attached to the leather strips were pieces of bone, glass, or lead.

The lictor's main target would be the accused back and extremities. Anything within reach of the flagrum resulted in multiple lacerations after striking the victim and being pulled through the targeted area leaving ribbons of torn, bleeding flesh. The victim suffered lacerations over the entire torso, extremities, and head, exposing the underlying musculature and viscera. The major vessels were protected by the ribs, thereby preventing death at the whipping post from exsanguinations. Agony was prolonged as a result. Scourging at the hands of the Roman lictors often left the victim a helpless mass of tissue, which could not be recognized as human. In the case of Jesus' scourging, He too was unrecognizable as a man. Isaiah wrote about Jesus' visage being disfigured (Isaiah 52:14).

The lictors, having witnessed the ongoing events as they unfolded, may have acted more aggressively than normal in an attempt to coerce an admission of guilt from Jesus. In some cases, "under the fury of the countless stripes, the victims sometimes sank, amidst screams, convulsive leaps, and distortions, into a senseless heap: sometimes died on the spot: sometimes were taken away in an unrecognizable mass of bleeding flesh, to find deliverance in death, from the inflammation and fever, sickness and shame." Jesus' silence denied Pilate a confession he may have been eager for. Venting their frustration after getting no response probably played a part in the severity. The other was probably due to the fact that Romans in general held a grudge against Jews and Jesus was a Jew.

According to one physician's description of a Roman scourging it is doubtful whether the Romans made any attempt to follow the Jewish law pertaining to whipping a person. The Jews had an ancient law prohibiting more than forty lashes. The Pharisees always made sure that the law was strictly kept. They insisted that only thirty-nine lashes be given. In the case of a miscount, they were assured of remaining within the law.

A Roman legionnaire stepped forward with the flagrum, or flagellus, in his hand. This flagrum a short whip consisting of several heavy, leather thongs with two small balls of lead attached near the ends of each. The heavy whip was brought down with full force again and again across Jesus' shoulders, back, and legs.

At first, the heavy thongs cut through the skin. Then, as the blows continued, they sliced deeper into the subcutaneous tissues. This initially produced an oozing of blood from the capillaries and veins of Jesus' skin, and eventually spurting of blood from arteries in the underlying muscles.

The balls of lead produced large deep bruises, which were broken open by subsequent blows. Subsequently, the skin of

Jesus' back hung in long ribbons. The entire area struck repeatedly by the flagrum left an unrecognizable mass of torn, bleeding tissue. Subsequent lashes tore away the flesh until they exposed the underlying bones. The beating was stopped when it was determined by the Centurion in charge that the prisoner was near death.

Blood spatter covered the whipping post, the lictors, and the immediate area. This is evidence indicating the severity of the scourging. The amount of blood loss was enormous, most probably leaving Jesus in a pre-shock state. "When one-third of the blood volume is lost rapidly, shock ensues. Death results from a rapid fifty percent blood loss." During the scourging the body's defense mechanisms began to react to the invading pieces of bone, glass, and metal lacerating the flesh. The blood vessels constricted while blood platelets began their clotting mechanism to reduce blood loss. This, in part, prevented any additional rapid blood loss which might have otherwise resulted in Jesus' death at the whipping post.

A trail of blood spatter from arterial bleeding, passively lost blood, and transfer blood in the form of Jesus' footprints or whatever He may have come into contact with could easily be traced from the whipping post as Jesus made the trek back to the Praetorium.

Before leaving the place of scourging Jesus was allowed to put on His own garment again. This soaked up the blood from His torn flesh. The only earthly possession He could claim as His own was now saturated with His blood. Consequently, the robe became a gambling prize at the place of crucifixion.

Following Jesus' scourging, the Romans, in order to shame

Him further, removed His garment, put a purple robe on Him and placed a crown of thorns on His head. The crown of thorns, made of thorns from a Zizyphus spina christi or Acacia niltotica twisted into the form of a mock laurel, not only inflicted additional injury and pain, but upon penetrating the vascular scalp it added to the already appreciable blood loss. The physical effects of the crowning with thorns would most likely cause "trigeminal neuralgia due to irritation of the ophthalmic branch of the trigeminal nerve and branches of the greater occipital nerves ... characterized by electric shock-like pains across the face."

The Romans made a sport of Him, mocking Him and striking Him about the face and head. This forced the thorns deeper into His scalp adding to His already traumatized and weakened condition.

Then the soldiers of the governor took Jesus into the Praetorium and gathered the whole garrison around Him. And they stripped Him and put a scarlet robe on Him. When they had twisted a crown of thorns, they put it on His head, and a reed in His right hand. And they bowed the knee before Him and mocked Him, saying, "Hail, King of the Jews!" Then they spat on Him, and took the reed and struck Him on the head. And when they had mocked Him, they took the robe off Him, put His own clothes on Him, and led Him away to be crucified. (Matthew 27:27-31)

The robe Jesus wore was removed. This re-opened the wounds wherever the robe may have had contact with His body, leaving them to again bleed freely.

Pilate's actions in this Roman trial and his verbal exchanges with the Jewish leaders are subjects of debate. What may appear at first glance to be the responses of a weak, not wanting to get involved, Roman prefect was probably not the case at all. Pilate was a political ruler as well as a military leader. He used his political influences, backed up by his militia, to satisfy his Jewish political allies while maintaining some semblance of order out of what could have easily resulted in utter chaos.

Pilate was known to be a cruel, hateful ruler. He knew how to appease the crowd while at the same time flex his military might over them. Pilate was not swayed by their biases. Jesus posed no threat to him. The only ones threatened were the ones wanting Jesus "out of the way." Since Romans despised the Jews, this may be the reason Pilate seemed somewhat sympathetic toward Jesus instead of exhibiting an aura of someone who was weak and easily influenced.

Throughout the morning someone in the crowd made comments such as, "If you release Him, you're no friend of Caesars." Pilate likely felt as though he was in a Him-or-me situation. If he released Jesus, what would become of his position as far as Rome was concerned? Pilate made the decision to save his career and at the same time made it look as if he would have nothing to do with Jesus by an open display of washed hands. Pilate's actions suggested he had no intentions of putting an innocent man to death and therefore, washed his hands saying "I am innocent of His blood."

The crowd's response was this: "Let His blood be on us."

Using their own customs, Pilate gave the Jews a choice: Jesus or Barabbas. They chose Barabbas.

"What then shall I do with Jesus?"

They said, "Let Him be crucified."

The crowd was given what they wanted. Jesus was led away to be crucified.

Reconstruction of the morning's events leads to this conclusion: innocence was suffering unjustly for the ones who were guilty. It was the only way to rid themselves of someone they did not want around.

The travesty of Jesus' trials brought out this opinion from a well-respected judge: "Jesus was judged before He was tried. He was charged and tried for three separate and distinct crimes. The Sanhedrin illegally convicted Him of blasphemy. Pilate refused to recognize this initial proceeding. Pilate twice acquitted Jesus of the charge of treason. He was charged with sedition before both Pilate and Herod but was acquitted by each. Yet Jesus was executed under a pretense he had been found guilty of treason. Threatened with possible loss of his position, Pilate chose to crucify Jesus as the easiest way to silence the angry priest."

At the end of what turned out to be several trials, both Jewish and Roman, we observed the following:

1. Initially, the hearings conducted by the Sanhedrin were at night and in secrecy. This is a direct violation of Jewish law (Mishnah Sanhedrin 4.1). The entire process was without any semblance of order.

2. Jesus stood alone, without the benefit of any representation or advice from council.

3. The involved parties appeared to have their minds set on ridding themselves of Jesus. The "charges" brought against Him were without merit.

4. Witnesses had been allowed, and even prompted, to

testify against Jesus without fear of being charged with perjury. Fabrications and circumstantial evidence without any basis of truth were admitted over truth.

5. The Roman "trials" appeared chaotic and full of indecisiveness. Too many questions were left unanswered. Pilate failed to grasp truth in his decision to proceed. He chose to heed a warning from his wife to have nothing to do with Jesus; signified his intent by washing his hands.

6. Political-type decisions were made to save face and appease the people.

7. Sentencing was without justification. Instead of being acquitted in a case without sufficient evidence to convict, Jesus was declared guilty and sentenced to death.

8. Punishment following declaration of guilt did not fit the accusations. An innocent man faced execution of a brutal nature at the hands of a foreign government.

4

Crucificxion

"If we are to be threatened with death, then we want to die in freedom; let the executioner, the shrouding of the head and the very name of the cross be banished from the body and life of Roman citizens, and from their thoughts, eyes and ears."

Cicero

On His way to Golgotha, Jesus was no longer able to stand erect due to His weakened physical condition. The scourging and everything else He had experienced up to that time drained His strength. This made the approximate half-mile journey on foot from the Fortress Antonia to Golgotha a slow one.

In addition, Jesus was forced to carry the patibulum (the cross member of the cross), which weighed in excess of fifty pounds; possibly as much as one hundred pounds. As a consequence,

He fell numerous times while carrying it. Since His hands were secured to the patibulum, it was not possible for Him to brace and prevent falling. This meant He would have fallen face first against the stone street. Any fall resulted in additional blunt force injuries to His face, torso, and extremities, with the possibility of internal injuries; compounded through compression from the load he carried. The only relief Jesus received was when Simon the Cyrenian was delegated to carry the patibulum the remainder of the way to Golgotha.

Jesus' body was lacerated and bleeding. He left a trail of blood as He traveled from Fortress Antonia to Golgotha. With each step transfer blood marked the trail in the form of Jesus' footprints. Passively lost blood dropped to the ground indicating the direction and slowness of His gait. Patterns of blood spatter depict the places He fell. The cross He carried was stained from its contact with Him.

The evidence thus far reveals the viciousness of the Roman's treatment of the condemned Jesus. It was not over. The worst was yet to come.

And He, bearing His cross, went out to a place called the Place of a Skull, which is called in Hebrew, Golgotha, where they crucified Him. (John 19:17-18a).

Execution by means of crucifixion was one of the most, if not the most, horrific way to be put to death. It has been identified as the most wretched of deaths. Crucifixion was a well-known form of capital punishment for slaves in the ancient world. Mass crucifixions during this same time period are well documented in literature on the subject.

Crucifixion was atrocious and barbaric. Even the most

notorious criminal would not deserve this type of inhumane treatment. "The crucifixion of a man is indeed irreconcilable with humanism of whatever hue. Every crucifixion casts doubt on man's claim to civilization ... When a man, no matter whom, is tortured and put to death in as cruel a manner as by crucifixion, it must lead to a crisis of human civilization."

The Romans flexed their might in ruling over their subjects. This was evident in the crucifixion of criminals and slaves who became victims of this Roman form of execution. Many procedural details involving crucifixion at the hands of the Romans correspond with details documented in the Gospels. They are listed as follows:

1. Scourging was ordered before execution. (Pilate ordered Jesus scourged. This order was carried out at the hands of Roman lictors before crucifixion.)
2. A tablet, identifying the reason for conviction, was hung around the neck. (A tablet for Jesus bore the inscription: Jesus of Nazareth, The King of the Jews. It was affixed to the cross in public view.)
3. A condemned person carried a patibulum (the transverse bar of the cross) to the place of execution. (Jesus carried His cross until Simon of Cyrene was forced to carry it for Him.)
4. Executioners were allowed to take possession of clothing worn by the condemned. (Roman soldiers gambled for Jesus' garment.)
5. Once at the site of execution the condemned was placed

on the ground and the arms/wrist affixed to the patibu-lum using ropes or nails. (Jesus was nailed to the cross.)

6. The condemned was then raised and the patibulum attached to a vertical pole already in place, then the feet of the condemned were affixed to the vertical pole in the same manner. (Jesus was suspended on the cross between Heaven and earth; His feet were nailed in place.) Archaeological evidence uncovered just north of Jerusalem in 1968 by V. Tzaferis supports the custom of using nails to affix condemned persons to crosses for the purpose of crucifixion. One discovery, identified as Jehohanan, was found to have markings on his radial bones and an iron nail through both heels.

7. An anaesthetizing drink was offered. (Jesus was offered this while on the cross but when He tasted it, He refused the drink.)

8. The lower legs of the condemned were broken to hasten death. (The soldier was directed to break the legs of the three condemned, but when he saw that Jesus was already dead, he pierced His side instead.) This is supported by the 1968 archeological discovery.

King David, the second king of Israel, wrote about the pangs of a crucifixion-type death more than one thousand years before Jesus' birth. What he described in *Psalms twenty-two* was a form of execution that would not exist for another five centuries.

• I am poured out like water. (Moment by moment vitality is stripped from the one crucified. Due to the tremendous strain placed on the body it gets weaker and weaker until finally overtaken by death.)

- All bones out of joint. (The body's weight and position on the cross results in the dislocation of joints.)

- My heart is like wax. It is melted in my body. (Physiologic changes in the person's organs begin as a result of blood loss. The body's attempt to compensate results in accumulation of blood to certain vital organs, which compromises their functional capacities. Blood and "water" flowed from the wound when the soldier pierced Jesus' side.)

- My strength is dried up like a potsherd. (The effect of bodily fluid and blood loss along causes dehydration, which is felt throughout the body.)

- My tongue cleaves to my jaws. (The tongue swells due to dehydration and adheres to the oral cavity. This is the reason Jesus said, "I thirst.")

- They pierced my hands and feet. (The hand and feet of condemned persons were secured to the cross with nails.)

- All my bones stare at me. (Lacerated flesh from scourging exposes the bones. Shamed from being stripped of clothing adds to one's dilemma.)

- They part my garments and cast lots. (Clothing of a person crucified was allowed to be distributed among Roman soldiers. The recipient was determined by means of a lottery.)

Witnessed Pain & Suffering:
Any witness to the agony and pain of individuals who have

been beaten, shot, stabbed, or involved in severe and fatal motor vehicle crashes is something a person will not soon forget.

Being with families as they are given news about the death of a cherished loved one and watching their countenance change upon hearing it is dreadful.

I can only imagine how horrible it was for Jesus' mother Mary, and the few friends accompanying her to Golgotha on the morning her son was crucified.

Sitting down, they kept watch over Him there. (Matthew 27:36)

A large crowd assembled on Golgotha for various reasons.

Members of the Sanhedrin were present to witness Jesus' crucifixion. These religious leaders took delight in joining the two thieves, who were condemned to die that day along with Jesus, in reviling and mocking Jesus. The actions of these pompous priests certainly were not representative of what good "religious" examples should portray.

Some of the Roman soldiers watched and kept order while others gambled for Jesus' clothing. The Romans appeared apathetic to the entire incident. They were present because of duty and were just following orders to carry out the execution of three condemned men. Later in the afternoon one soldier realized the significance of Jesus death. Historical records identify him as Longinus. He said, "Truly this Man was the Son of God" (*Mark 15:39*).

Relatives present included His mother and an aunt. Others included followers, friends, and one of the twelve disciples. The only direct dialogue between Jesus and persons witnessing the event was when He spoke to His mother and the disciple (*John 19:26-27*). Obviously, Jesus could see and feel their heartache.

A mixture of emotions could be detected from many faces around the cross. Some of them smiled and even laughed while others exhibited a caring spirit and grieved at the sight of Jesus hanging on the cross. Each was eyewitness to His torture and last few hours of life. Witnessing an incident like this would be horrible enough. Trying to imagine having to personally experience this kind of torture and death is beyond comprehension.

Several factors influenced Jesus' physical ability to tolerate the type of battering and torture. Among them are the following:

1. Jesus was the age of thirty-three years when He was crucified. He would be considered in the prime of His life.
2. Jesus worked as a carpenter up until about the time He began His ministry at the age of thirty. In terms of physique, Jesus would be considered lean and muscular.
3. During Jesus' three-and-one-half-year ministry the only method of land travel He took was on foot. Barring any natural disease process, He would be considered in great cardiovascular condition.

Jesus was God-in-flesh. Although Jesus' body was mortal, as Christ His mind is immortal. While on earth in a mortal body Jesus' thought process was not anything like other mortals.

Even for someone in prime physical condition, the amount of pain and suffering Jesus endured is beyond comprehension. The magnitude of all he endured would have quickly taken its toll on Him. Pain and suffering of this nature would have broken down and made weaklings out of the strongest humans. If you will stop for a moment and think about every hurt you have ever experienced, every pain you have ever had in your body, every traumatic moment in your life up until now, and every worry that may have seized your mind, you may understand a small part of what Jesus went through. Now multiply all of your hurts, pains, traumatic moments and worries by the number of humans ever to live on this earth, along with those yet to be born, place it on Jesus and you may get a better understanding of how greatly He suffered. Include eternal judgment for humanity's sin and you have an idea about how much pain and suffering Jesus accepted for our sakes in less than a twenty-four-hour period on the day He was crucified.

On the cross, Jesus' exposed body was suspended between heaven and earth in the heat of the noon sun. His constrained and fixed position on the cross caused acute pain to escalate as the moments passed. Where He might find some relief by trying to move, any movement only increased the pain associated with the trauma He sustained. The sensitive nerves in His hands and feet were crushed and mutilated from the nails having pierced them. Any movement causing contact with the median nerve in the hands and the medial plantar nerve in the feet sent searing, unrelenting pain like lightning through His arms and legs. Inflammation of the wounds set in. His hands and feet swelled and ached. Movement proved extremely painful. Every time Jesus pushed up on the cross to breathe, pain rushed up from His feet. His raw, exposed back rubbed against the cross. Splinters from the wooden cross jabbed and pierced His flesh. His entire body writhed in agony. His injured body was covered in drying and clotted blood. Every movement opened some wound, which bled freely until it again clotted and dried.

Jesus' body underwent physiologic changes to compensate for the amount of blood loss. His heart rate increased. His blood pressure dropped. His reduced blood volume resulted in redistribution of blood to His vital organs leaving an insufficient volume of circulatory blood to other parts of His body.

Jesus experienced exhaustion and dehydration. The only thing He asked for was a drink when He said, "I thirst." Everything else was for the benefit of others. By His actions Jesus assumed responsibility and paid the penalty for sin in our stead.

Jesus' ability to endure the types of extreme torture He had to go through, as described in the Gospels, is a subject of debate.

1. In the Garden of Gethsemane Jesus went through the described combat of will. His body responded to this stressful condition and began to sweat blood.

2. Jesus was betrayed and unlawfully arrested while enduring emotional shame and disgrace at the hands of the Sanhedrin and Roman soldiers.

3. A crown of thorns was placed upon His head. He was beaten and scourged until he no longer appeared to be human. The severe blood loss caused His body to go into traumatic shock.

4. Jesus was forced to bear a seventy-five-pound patibulum on the way to the place of His execution, falling under its weight and, more probable than not, sustained additional injuries from blunt trauma.

5. Upon reaching Golgotha, His hands and feet were nailed to a cross. On the cross, Jesus suffered like no other human ever suffered, including eternal death for humanity.

What are the possibilities of any mortal enduring the many types of tortuous treatment Jesus suffered and still have the human capacity, both physically and emotionally, to bear the weight of a cross and remain alive for a few hours after being nailed to a cross? One answer is found in the Roman history of crucifixion. Scourging before crucifixion was common practice. Following scourging, condemned persons were nailed to a cross where they would remain sometimes as long as nine days before succumbing to the power of death.

Even so, in addition to scourging and being nailed to a cross the treatment Jesus endured appears to have been more severe than the routine procedure of a Roman crucifixion. The effect of everything Jesus went through seems disproportionate when compared to the routine. Jesus did not last days on the cross. His death occurred within a few hours of being scourged and nailed to the cross.

When we observe Jesus' ability to tolerate this type of torturous treatment we tend to think of Jesus in finite terms. Like you and me, Jesus was human. The five senses of sight, hearing, touch, taste and smell were as significant in His life as they are in ours. His humanity is not in dispute. Jesus' human side was relegated to time and space; something He was not used to. Therefore, when the writer of Hebrews stated that Jesus "*endured*" the cross (*Hebrews 12:2*), it signified that for a length of time a load existed for Him to bear. The weight of the world was placed upon Him all at once. Every moment of the three hours following the time He was fastened to the cross in all probability seemed like an eternity—especially for someone who was innocent of any wrongdoing. The end must have seemed endless.

Nevertheless, when considering what Jesus had to go through in the final hours of His earthly life, we must also consider *who* Jesus is. In body, Jesus was human—as human as you or me. In mind, Jesus is God. Although Jesus felt pain and weakness in His finite body, suffered as anyone else would under the same circumstances, and knew that His correspondence with the environment was about to cease, His will to survive and complete His mission drove Him onward. When human strength failed, His divine infiniteness kept Him alive to implement the work He came here to do to absolute success.

John Gill (1697-1771) described Jesus' final hours on the cross. When Jesus was made sin and a curse it was tantamount to an eternal death, or the suffering of the wicked in hell. For though the two kinds of suffering differ as to circumstances of time and place, the persons were different. We have the finite and the infinite. In essence of these sufferings, they were the same. Eternal death consists in two things: Punishment in the form of deprivation, and punishment in the form of actual affliction. Punishment in the form of deprivation lies in an eternal separation from God, or a deprivation of his presence forever. Punishment in the form of affliction lies in an everlasting affliction in the everlasting fire of God's wrath. Christ endured what was answerable to both. Eternity is not the essence of punishment but is consequent of the fact that the sufferer cannot all at once bear the whole; being finite as sinful man is finite. Since it cannot be borne all at once it is continued ad infinitum. Nevertheless Christ, being infinite, was able to bear the whole at once and the infinity of his Person abundantly compensates for the eternity of the punishment.

Berated and scoffed, Jesus looked around at the two thieves who were crucified with him; the Roman soldiers guarding the three crosses; members of the Sanhedrin waiting around for His final moment that they might claim victory; and the few faithful followers determined to stay by Him until the end.

Even though Satan had "bruised his heel," with every ounce of strength Jesus, probably thinking "just a few more moments," determined to overcome the last obstacle—the enemy of Death. At the moment of death Jesus cried with a loud voice "Teleiotes." It is finished!

5

Cause of Death

Ordinarily when someone dies following some traumatic incident, a post-mortem examination and autopsy are performed. The findings are posted with a definitive cause of death based on the results of the examination.

If a body is not available for examination, or for some reason an autopsy is not possible—either permission is denied for personal or religious reasons—then circumstances leading up to the time of death, what transpired at the moment of death and shortly thereafter, must be considered to establish a probable cause of death. This is identified as a psychological autopsy.

Everything Jesus suffered from Gethsemane to Golgotha must be considered when attempting to establish His cause of death. Emotional and physical traumas are considered in the final diagnoses.

Physical trauma is easily recognized based on the type of physical injury sustained, whether the injury is inflicted by blunt force (beating), sharp force (cutting), or other means whereby an injury pattern is identifiable exists.

Determining the influence of emotional trauma on a person's demise may be possible by learning what events lead up to the time of death. How these events might have affected the cause may be evident based on the individual's age, physical condition, and type of physical trauma observed to the body.

Events to consider in the death of Jesus Christ are:

1. Events of Gethsemane.
2. Events of the Sanhedrin trial.
3. Events of the Roman trial.
4. Events of Golgotha.

Events of Gethsemane:

1. Jesus agonized about His becoming accountable for humanity's sin before yielding to His Father's will. The guilt of every human to ever live fell on Him. Anxiety of the impending Roman scourge and crucifixion weighed heavy on His mind. The stress of this event took an emotional and physical toll on His body. Blood loss through His sweat glands and sensitive-to-touch skin came as a result.
2. Weak and bloody from three hours of agonizing prayer, Jesus expressed concern to the disciples about their sleeping instead of keeping watch for Him. This, in all probability, added to His dejection.

During this same time, Jesus was betrayed by one of the twelve disciples. Judas arrived at the garden with armed men to affect His arrest.

Events of the Sanhedrin trial:

1. Jesus faced the mockery of a trial among people who had already decided His fate. The charges against Him were without merit. Jesus was unjustly convicted and condemned to death.
2. Jesus suffered physical and emotional abuse at the hands of those present.

Events of the Roman trial:

1. Jesus was scourged by Roman lictors, which resulted in a large quantity of blood loss and bodily disfigurement.
2. Roman soldiers humiliated Jesus when they placed a purple robe on Him, then began mocking, pulling out His beard, beating, and spitting on Him in addition to forcing a crown of thorns down onto His head.
3. Knowing Jesus was innocent, Pilate wanted to release Him but the crowd repeatedly chanted, "crucify Him." Not one person came forward to testify on His behalf. Jesus stood all alone, despised and dejected.
4. Jesus was forced to carry His cross to the place of execution, in spite of His weakness from injury, lack of sleep and dehydration.

Events of Golgotha:

1. Jesus' hands and feet were secured to the cross with long nails.
2. Jesus suffered further ridicule from some of the ones present.
3. The heat of the noonday sun further complicated His already dehydrated condition.
4. The position of His body on the cross made it difficult to breathe.
5. Movement on the cross exacerbated the already excruciating pain Jesus suffered from His injuries.
6. Physiologic changes in His body due to everything He had been through caused additional suffering and finally death within a few hours.

As a sacrificial lamb, Jesus became the thing He abhorred —sin.

In addition, Jesus was forced to go without sleep, food, and water from the time He left for Gethsemane until His death the following afternoon.

Based on available evidence opinions about an exact cause of death differ.

In *Death by Crucifixion,* the authors opine that it appears likely that the mechanism of death in crucifixion was suffocation. They suggest the chain of events which ultimately led to suffocation support their opinion. The weight of the body results in the arms being pulled upward. This caused the intercostal and pectoral muscles to be stretched. Furthermore, movement of these muscles was opposed by the weight of the body. With the muscles of respiration thus stretched, the respiratory bellows became relatively fixed. As dyspnea developed and pain in the wrists and arms increased, Jesus was forced to raise His body, thereby transferring His weight to the feet. Respirations were easier, but the weight of the body was then exerted on the feet and pain in the feet and legs increased. When the pain became unbearable Jesus again slumped down with the weight His body pulled on the wrists and again stretching the intercostal muscles. Thus, He alternated between lifting his body in order to breathe and slumped down to relieve pain in the feet. Eventually, He became exhausted and possibly lapsed into unconsciousness so that He could no longer lift His body. In this position, with the respiratory muscles essentially paralyzed, Jesus suffocated and died.

The Chief Medical Examiner of Rockland County, New York, presents the following: "Could a person in a state of traumatic and hypovolemic shock who had undergone severe anxiety to a point of hematidrosis, had been brutally scourged with a flagrum, suffered trigeminal neuralgia from the crowning with thorns, stumbled and fell for a half mile carrying a fifty pound cross part of the way, then nailed through the hand and feet with large spike-like nails and suspended on a cross be able to repeatedly push and pull themselves up against the spike-like nails in their swollen, exquisitely tender hands and feet in order to breathe over a period of several hours? I don't think so...all of these factors revealed the cause of death in crucifixion to be due to traumatic and hypovolemic shock."

The opinion of other physicians indicates the cause of Jesus' death may have been a combination of factors. "The fact that Jesus cried out in a loud voice and then bowed His head and died suggests the possibility of a catastrophic terminal event.

One popular explanation has been that Jesus died of cardiac rupture...Jesus' death may have been hastened simply by His state of exhaustion and by the severity of the scourging, with its resultant blood loss and pre-shock state. The fact that He could not carry His patibulum supports this interpretation. The actual cause of Jesus' death, like that of other crucified victims, may have been multifactorial and related primarily to hypovolemic shock, exhaustion asphyxia, and perhaps acute heart failure."

Another physician stated, "On the cross the workload of the heart was greatly increased due to multiple factors, but primarily the increased effort necessary to breathe. This resulted in a rupture of the free wall of the heart, which caused Jesus to cry out in a loud voice and suddenly die. This cause of death is confirmed for us by the sword pierced to the side, which resulted in the flow of blood and water."

Other possible causes of death are identified as: "cardiac rupture or cardio respiratory failure, associated hypovolemia, hyperemia, and an altered coagulated state."

A more recent article attributes the cause of Jesus' death to venous thromboembolism (blood clot). In the author's opinion the evidence fits well with Jesus' condition and in all likelihood was the major cause of death by crucifixion.

Pulmonary embolism is a common cause of death among victims known to suffer from combined effects of trauma, immobilization, and dehydration. Immobility following trauma is a major factor in the formation of embolisms. Findings of death by pulmonary embolism are well documented among the traumatic deaths investigated by medical examiners during autopsy. Without an autopsy many of these embolism deaths would likely be attributed to inadvertent misdiagnosed etiology. Therefore, without an autopsy of Jesus' body we have to rely on the few sketchy accounts of events surrounding His death.

The result of the soldier piercing Jesus' side at the time of His death provides evidence to substantiate a possible definitive cause of death. The apostle John documented the event, which described fluid coming out of the wound as an exuding of blood and water. It is feasible that an escape of watery fluid from the pericardial sac around the heart along with blood from the interior of the heart is conclusive post-mortem evidence that Jesus died of heart failure due to shock and constriction of the heart by fluid in the pericardium and not the supposed crucifixion death by suffocation.

Matteo Bevilacqua, Giulio Fanti, and Michele D'Areinzo wrote an article, The Causes of Jesus' Death in the Light of the Holy Bible and the Turin Shroud, published on April 11, 2017, in *Open Journal of Trauma*. In this article, they state, "It is likely that the blood that came out of the pericardium wound was mixed with the blood contained in the pleura. This does not conflict with the theological thought of the Christian church ... From the discussion of the various hypotheses, we deduce that the most likely immediate cause of Jesus' death was myocardial infarction complicated by heart rupture."

The trail of Jesus' blood from Gethsemane to Golgotha is a constant reminder of the non-stop aggression Jesus encountered during the final hours of His life on earth.

The last thing to scrutinize in this issue is responsibility. Who or what is accountable for Jesus' death?

6

Accountability

Through the years many opinions have been offered about accountability relating to the death of Jesus Christ.

Anti-Semitism is prevalent in many opinions making it easy to point the finger at the Jews. Many who express this opinion base their views, in part, on the passage of Scripture, *"He came to His own, and His own did not receive Him."* Members of the Jewish Sanhedrin consorted and condemned Jesus before delivering Him to the Romans for trial and execution.

Blame has been placed on Pontius Pilate and the Romans. Rome existed as the governing authority in Israel at the time of Jesus' death. Pontius Pilate single-handedly had the authority to release Jesus. Expressed opinions place accountability on Pilate who could have released Jesus if he entertained the choice. Instead, Pilate sent Jesus to His death and as a symbol of absolving

62

himself washed his hands of Jesus saying, "I am innocent of this man's blood."

Other persons are identified in this matter who in some manner expressed a desire to have Jesus eliminated. Some of these individuals intended to see that this was achieved by whatever means possible.

They had motive.

Motive is an underlying reason someone may wish harmful action to occur to another individual. A person is usually identified as a suspect after such motive is established, whether expressed by word or deed.

Motivation may be established by learning the answers to various questions. In this case, one of the most important questions to ask is: Who would benefit the most from Jesus' death?

Identifying logical suspects based on motive in this matter is the easy part. Many people ostracized Jesus for a multitude of reasons. Asking a few questions and performing a little research and follow-up to discover who may have wanted Jesus to die is not a complicated matter.

Today, if the Court intended to indict someone for an incident, the only evidence necessary is enough to establish probable cause. To bring about a conviction in front of a jury of their peers the Court must have enough evidence to prove guilt beyond reasonable doubt.

Consider the individuals who may have had a motive in the case against Jesus. Many names come to mind when thinking about culpability. Some individuals we may think of are those who were closest to Jesus during His ministerial years. Others are known to have harbored animosity toward Him for personal reasons. In addition, we should identify and consider the ones who would benefit the most from Jesus' death.

Judas Iscariot

One of the obvious suspects to consider is Judas Iscariot. Judas was one of twelve fortunate individuals chosen by Jesus to support Him in ministerial duties.

Judas Iscariot expressed intent by his actions when he met with the high priests and elders for the purpose of striking a deal with them before betraying Jesus.

What was Judas' motive for wanting Jesus arrested? If it was for fortune, Judas walked away with thirty pieces of silver. If it was for fame, he got his wish. This collusion between Judas and the high priests and elders made him notorious. He may have hoped the events following the betrayal would force Jesus to reveal His identity and assume the role of a political-military messiah, lead to an overthrow of Roman rule while taking over David's throne as King of the Jews. This would have been favorable for Judas since he was one of the twelve chosen disciples.

Judas may have hoped for acceptance by the Jewish religious leaders and the possibility of being given some prestigious position among them.

Whatever Judas' motive could have been is mere speculation. A motive was probably garnered from his conceit.

One thing is certain. Judas made a deal with the Jewish religious leaders to betray Jesus for thirty pieces of silver. Then, whether a result of a changed mind due to a guilty conscious or the result not being what he expected would happen, he tried to return the money. Judas knew Jesus was as good as dead when Jesus was turned over to Pilate. There was no turning back at that point. Within hours, Judas' body was found hanged outside the city. The manner of death was suicide.

The Jews

What about the motives of the Jews? Is an entire nation to be blamed for the motives and actions of a few Jews who may have wanted Jesus to die? Jesus had many followers among the Jewish nation of Israel. His presence among them brought many of them relief. Jesus catered to their needs. The ones who seemed to harbor the most animosity against Jesus were legal and religious leaders; more specifically the Sanhedrin.

"Many from the crowd, when they heard this saying, said, 'Truly this is the Prophet.' Others said, 'This is the Christ.' Others said, 'Will the Christ come out of Galilee?' ... So the was a division among the people because of Him. Now some of them wanted to take Him, but no one laid hands on Him. Then the officers came to the chief priest and Pharisees, who said to them, 'Why have you not brought Him?' The officers answered, 'No man ever spoke like this Man!' Then the Pharisees answered them, 'Are you also deceived? Have any of the rulers or the Pharisees believed in Him? But this crowd that does not know the law is accursed.'" (John 7:40-49)

The Sanhedrin sought ways to justify an arrest and support a death sentence. From all indication it seems that the only real *problem* they had with Jesus was His identity and truthfulness. At times their anger reached such proportions that they were ready to stone Him because some of His statements or conduct did not coincide with their tradition or part of Jewish Law. In addition, they questioned Jesus' authority and could not dismiss His claims to be God's Son, which made Him equal with God.

Instead of accepting Jesus, many rejected Him and dismissed

His teaching as heresy. In their eyes Jesus was not *good enough* to be their long-awaited Messiah. He was not what they expected in a Messiah. They may have disliked Him because of His appearance. They may have disliked His demeanor. In their opinion, Jesus may not have had the proper standing to be named Messiah and therefore, should not be allowed to rule over them. Jesus was often referred to as "the son of a carpenter." They despised Jesus for keeping company with sinners. They may have decided that Jesus simply did not fit in with any of their future plans.

At times before, these same religious leaders attempted to take Jesus because of His words or actions. The first occurred after Jesus gave His oration in the synagogue as recorded in Luke 4:16-30. Jesus was taken outside the city of Jerusalem where they intended to throw Him from a cliff, but He was able to slip away and escape. The Apostle John recorded the Pharisees saying that Jesus testified of Himself and His "testimony" was not true (John 8:13-14). As recorded by John, Jesus said, "*Now you seek to kill Me.*" They even accused His mother of fornication. They picked up stones to stone Him but He hid and was able to get away. Subsequent to that the Pharisees said Jesus was "not of God" (John 9:16) because He did not "keep the Sabbath" since He performed a miracle on the Sabbath day. A similar incident when they picked up stones to stone Him was recorded in John 10:30-39 after Jesus made the statement, "*I and My Father are one.*"

The only motive evident following these incidents is their expressed displeasure in His claim to be the Son of God. More probable than not some members of the Sanhedrin had already made up their minds about Jesus' guilt long before He ever appeared for trial. This was especially true for the High Priest, Caiaphas.

The assembly outside the Praetorium waiting on Pilate responded *"His blood be on us and on our children."* Pilate washed his hands and declared he would not be held responsible for taking the blood of an innocent man. Therefore, accountability seems self-imposed based on the bloodthirsty crowd's response.

Caiaphas

Caiaphas served as High Priest for eighteen years (A.D. 18-36). He was the second most powerful dignitary in Judea under Pontius Pilate. Caiaphas' actions relate to political prowess. His presence among the Romans and his own nation was like a lion seeking prey to consume for his own political enhancement.

Caiaphas consorted with his father-in-law Annas, and other members of the Sanhedrin to falsely condemn Jesus and convince the Jewish people that Jesus was the chosen one who should die.

Then the chief priest, the scribes, and the elders of the people assembled at the palace of the high priest, who was called Caiaphas, and plotted to take Jesus by trickery and kill Him. (Matthew 26:3-4)

Now it was Caiaphas who advised the Jews that it was expedient that one man should die for the people. (John 18:14).

The one thing many agree to as being the climactic act to further their hatred for Jesus was when Jesus drove out the moneychangers from the Temple in Jerusalem. "At a time of high tension such as the Passover festival, it is likely that any subversive action in the Temple—even action of a symbolic nature—would provoke a strong response from high priests and Roman officials. It did."

Therefore, because of his role in the events leading to Jesus' demise, Caiaphas is considered a prime suspect.

The Romans

The only motive evident among the Romans was their hatred of the Jews.

Pontius Pilate gave no indication during Jesus' appearance before him of any animosity against Jesus. He was willing to release Jesus and gave the impression by his actions (washing his hands of the matter) and statements (i.e., "I find no fault in this man.") that he did not want anything to do with His demise. The only indication of anything to be considered is the accusation made by Caiaphas implicating Jesus of insurrection and the statement, "If you let Him go you are no friend of Caesars."

Last Will and Testament:

In the seventeenth chapter of John's gospel, we find the intercessory prayer Jesus prayed a short while before His arrest in Gethsemane. This prayer may be construed as the providential expression of the Last Will and Testament of the Lord Jesus Christ.

Jesus Christ and eternal life are synonymous. Accompanying eternal life is the inheritance comprised of everything Heaven has to offer us.

In John seventeen, Jesus acknowledged His right to allocate the gift of eternal life to humanity with the stipulation we meet this requirement: believe His death was for humanity and accept Him as Savior.

For this Last Will and Testament to be established and enforceable Jesus would have to die.

For this reason, He is the Mediator of the new covenant, by means of death, for the redemption of the transgressions under the first covenant, that those who are called may receive the promise of the eternal inheritance. For where there is a testament, there must also of necessity be the death of the testator. For a testament is in force after men are dead. (Hebrews 9:15-17)

In this ordinary manner Jesus named His joint-heirs. Through the extraordinary expression of love by His death on a cross, thereby accepting our accountability, we are afforded the opportunity to meet the requirement.

In Him also we have obtained an inheritance, being predestined according to the purpose of Him who works all things according to the counsel of His will. (Ephesians 1:11)

The word inheritance used by the Apostle Paul in the above verse is the Greek word "kleroo," meaning to assign a portion; to be made heritable; designated as a heritage. Referring to the inheritance the Apostle Peter used the word "reserved" (1 Peter 1:4), translated from the Greek "tereo," meaning contracted and held in reserve.

The prospect of life over death after Adam sinned in the Garden of Eden depended on one thing—a sacrifice worthy of full pardon for sin must be offered. When sin was introduced and Adam succumbed to temptation, the Second Law of Thermodynamics assumed control. This led humanity on a downward spiral from order to disorder, until one day they would reach a state of complete randomness and finally, death. The only worthy sacrifice capable of redeeming humanity from their natural digression and deterioration was Jesus Christ.

Conclusion

An examination into the life of Jesus Christ provides insight into the torturous death He suffered on the cross. Jesus' life was exemplary in nature. He was the perfect example to follow. Everything He did and said benefited others.

Jesus' actions exemplified love in its purest form. He denied himself many luxuries of daily life to provide for others. What was in His power to do, He did for the asking.

With open arms and kind words Jesus beckoned everyone to come to Him. Instead, the ones He had come to help turned their backs on Him. Instead of acceptance, He was openly despised and rejected by His own people.

Although Jesus had the power to save humanity from sin's dominating power over them, He was not one to force Himself on anyone. Freedom of choice is left up to the individual. Every person has the right to decide for their own selves whether or not to believe in what Jesus provided by His sacrificial death and accept Him as Savior.

Was it Judas Iscariot, the Jews, Caiaphas, or the Romans who gained the most benefit from Jesus' death? It was neither and it was all of the above. Humanity had the most to gain when Jesus died. Therefore, humanity's nature as a result of sin's introduction is ultimately responsible for Jesus' death on the cross.

Jesus overcame the temptation to come down from the cross and accepted what was before Him. Even the taunts of the ones at the cross were not enough to persuade Jesus to abandon His mission. Jesus would not allow accountability to be placed on humanity. Instead, He asked that forgiveness be granted. Jesus willingly gave His life for the very ones who at the same time wanted to take it from Him as though He were the enemy. Jesus willingly took humanity's sin upon His own body. That was evident by what occurred on Golgotha. As a result of Jesus' willingness to take humanity's sin and His ability to bear them in humanity's stead, the one-and-only-worthy sacrifice offered for sin was successful.

What took place on Golgotha was not the end of life. It was the end of the beginning of life for you and me.

Jesus' body was removed from the blood-stained cross on Golgotha and placed in a tomb belonging to Joseph of Arimathea. The tomb was sealed with a stone rolled across its entrance.

Pilate commissioned soldiers to guard the tomb at the request of the Jewish leaders.

After three days, the stone covering the entrance was discovered out of position; the tomb's interior exposed.

The soldiers commissioned to guard the tomb had deserted their post.

Jesus was gone.

About the Author

Steve Rush is an award-winning author whose experience includes tenure as homicide detective and chief forensic investigator for a national consulting firm. He was once hailed as "The best forensic investigator in the United States" by the late Joseph L. Burton, M.D, under whom he mastered his skills, and investigated many deaths alongside Dr. Jan Garavaglia of *Dr. G. Medical Examiner* fame. Steve has investigated 900+ death scenes and taught classes related to death investigation. His specialties include injury causation, blood spatter analysis, occupant kinematics, and recovery of human skeletal remains.

Steve's book *Kill Your Characters; Crime Scene Tips for Writers*, published by Genius Books on June 7, 2022, was named finalist in the 2023 Silver Falchion Award for Best Nonfiction and Honorable Mention in the 2023 Readers' Favorite Awards. Steve won joint first prize in the 2020 Chillzee KiMo T-E-N Contest and longlisted in the 2022 Page Turner Awards.

He lives in Metropolitan Atlanta, Georgia, with his wife Sharon.

Visit his website at: https://www.steverush.org/

Bibliography

Bevilacqua, Matteo; Fanti ,Giulio; and D'Areinzo, Michele; *The Causes of Jesus' Death in the Light of the Holy Bible and the Turin Shroud*, Open Journal of Trauma, April 11, 2017.

Ball, David A., M.D., *The Crucifixion of a Man Called Jesus*, Journal of Mississippi State Medical Association, March 1989.

Bev,l, Tom and Gardner, Ross M., *Bloodstain Pattern Analysis*, Second Edition, CRC Press, Copyright 2002.

Bishop, Jim, *The Day Christ Died*, Harper & Brothers, N.Y., Copyright 1957.

Brenner, B., *Did Jesus Christ Die of Pulmonary Embolism?* Journal of Thrombosis and Haemostasis, Copyright 2005.

Charlesworth, J.H., *Jesus and Jehohanan: An archaeological Note on Crucifixion*, Expository Times, February 1973.

Davis, C. Truman, M.D., *The Crucifixion of Jesus: The Passion of Christ from a Medical Point of View*, Arizona Medicine, 1965.

DePasquale, N.P. and Burch, G.E., *Death By Crucifixion*, American Heart Journal 66(3), Copyright 1963.

Edwards, William D. MD.; Gabel, Wesley J. MDiv.; Hosmer, Floyd E. MS, *On the Physical Death of Jesus Christ*, JAMA 1986.

Fogle, Hon. Harry, *The Trial of Jesus*, Jurisdictional Foundation, Inc., Copyright 2000.

Geikie, Cunningham, *The Life And Words Of Christ*, Copyright 1879.

Gill, John, *A Body of Divinity*, Grand Rapids, Baker reprint, Vol. I.

Harrub, Brad, Ph.D. and Thompson, Bert, Ph.D., *An Examination of the Medical Evidence for the Physical Death of Christ, Reason & Revelation*, Apologetics Press, Inc., Copyright 2002.

Hodges, Andrew G., M.D., Jesus: *An Interview Across Time*, Kregal Publications, Copyright 1986.

Josephus, Flavius, *Antiquities Book 18*.

Krejcir, Richard J., *The Crucifixion of Jesus*, Copyright 1982.

Linder, Doug, *The Trial of Jesus: An Account*, Copyright 2002.

Mason, J.K., M.D., *The Pathology of Trauma*, Edward Arnold, Copyright 1993.

Ruch, Theodore C., *Physiology and Biophysics*, Copyright 1965.

Spitz, Werner C., M.D., *Medicolegal Investigation of Death*, Third Edition, Charles Thomas Publisher, Copyright 1993.

Spurgeon, C.H., *The Agony In Gethsemane*, 1894.

Tacitus, *Annals*, 15.44.

Vonderahe, A.R., *New Scholasticism*, Copyright 1944.

Weber, Hans-Ruedi, *The Cross*, Kreuz Verlag Stuttgart, Copyright 1975, English Translation, SPCK, London, Copyright 1979.

Wonder, A.Y., *Blood Dynamics*, Academic Press, Copyright 2001.

Zugibe, Frederick T., M.D., Ph.D., *A Forensic Way of the Cross*, Copyright 2000.

Glossary

Asphyxia – a condition due to lack of oxygen resulting in cessation of life.

Dyspnea – having difficulty or labored breathing.

Embolism – sudden blockage of an artery by a clot or other foreign matter obstructing circulation of blood.

Etiology – cause of disease or disorder.

Hematidrosis – a process by which tiny capillaries in the sweat glands rupture resulting in bloody sweat.

Hyperemia – excessive blood in a part; engorgement.

Hypothesis – an assumption used to test logic.

Hypovolemia – abnormal decrease in circulation of blood in the body.

Neuropathology – the study of disease process in the brain.

Ophthalmic – pertaining to the eye.

Patibulum – a long beam of wood, usually weighing fifty- to seventy-five pounds, used as a cross member of a cross.

Pericardium – the fibro-serous sac surrounding the heart.

Praetorium – the ancient palace of a Roman magistrate.

Pulmonary – of or pertaining to the lungs.

Trigeminal neuralgia – excruciating episodic pain in the

area of the trigeminal nerve extending across the face to the
base of the brain.

www.ingramcontent.com/pod-product-compliance
Lightning Source LLC
Chambersburg PA
CBHW061439160726
47995CB00003B/955